Fossil News

The Journal of Avocational Paleontology

Vol. 22.2— Summer 2019

FRONT COVER: *Dilophosaurus*, an Early Jurassic dinosaur (the "two-crested lizard") first discovered in Arizona in 1940. *Dilophosaurus* was one of the earliest large predatory dinosaurs, and its fingers and toes bore impressive claws. FACING PAGE: Smilodont skull, based on a display at the Denver Museum of Nature and Science. BACK COVER: Reconstruction of a nautiloid cephalopod based on a reconstruction at the Natural History Museum of San Diego, California; and *Albertosaurus* skull, drawn from an exhibit at the Denver Museum of Nature and Science.

Contributors To This Issue

DAVID CORDIE (writer) is a recent graduate of the PhD program in geosciences at the University of Wisconsin-Milwaukee. While studying for his undergraduate degree in biology, David performed ecological studies on reefs from the Cayman Islands during a school trip. Later, he incorporated these interests into his studies on evolution and the fossil record. Currently he studies Cambrian ecology, reefs, systematics, and anything involving invertebrate fossils. He lives in Milwaukee, Wisconsin, and works part-time as a fossil preparator.

ENRICO BONINO (writer) received his degree in Geology from the Department of Earth Sciences at the University of Genoa. He currently lives in Belgium where he works as a cartographer specialized in the use of satellite-imaging and geographic information systems. He is the author of *The Back to the Past Museum Guide to Trilobites* (with Carlo Kier), and recently published *Spiagge Cambriane: Meduse e Tapetti Algali* ("Cambrian Shores: Jellyfish and Algal Mats"), excerpted in English in this issue.

NANDO MUSSMARA (writer) is the founder of *PaleoNature: The Multilingual Natural History Magazine* and a long-time fossil collector, science writer, and photographer. He has been a Contributing Editor to *Fossil News* since 2016. Visit PaleoNature at www.paleonature.org.

YASMANI CEBALLOS IZQUIERDO (writer) is a computer engineer and freelance writer. He cofounded the Cuban Digital Library of Geosciences eight years ago and regularly writes about science and paleontology for the Library as well as for other publications. He can be reached at yasmaniceballos@gmail.com, or see his work online at www.redciencia.cu/geobiblio/inicio.html.

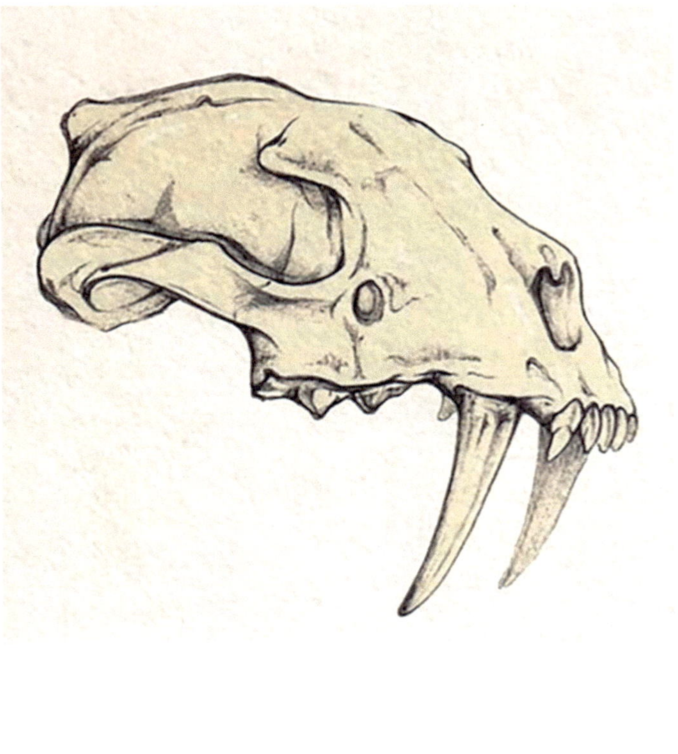

HOW TO SUBSCRIBE: Domestic US rates: $50.00 USD for four quarterly issues, payable via PayPal or Zelle (both via FossilNews@FourCatsPress.com) or Venmo (@FourCatsPress). Outside the U.S., please contact us for a quote. Single issues: $15.00 USD (plus postage if outside the US). Complete info at: tinyurl.com/fnsubscribe.

DISPLAY ADVERTISING: Subscribers receive 10% off display advertising. Our current "Rate Sheet & Specifications" are online at: tinyurl.com/fnadvertise.

SUBMISSION GUIDELINES: *Fossil News* welcomes submissions from all those with an interest in paleontology and fossil collecting. Submissions and queries to: FossilNews@FourCatsPress.com.

HOW TO SUBMIT YOUR WORK: Please read our guidelines (tinyurl.com/fnguidelines) or contact us for details.

BOOK REVIEWS: *Fossil News* does not accept unsolicited or previously published reviews. It's always best to query first. If possible, send clips of your published work.

EDITING AND REVISION: *Fossil News* reserves the right to edit submissions for length, grammar, punctuation, structure, and style.

PAYMENT: We offer contributors a complimentary copy of the issue in which their work appears and a choice of (a) two insertions of a display ad (up to one-half page) or (b) 50% off an annual or gift subscription.

Fossil News
The Journal of Avocational Paleontology
FossilNews@FourCatsPress.com
www.FossilNews.org
Summer 2019: 978-1-7320442-7-2

Featured Paleoartist

Jimi CATANZARO

Las Vegas native Jimi Catanzaro is only 30, but he's already in the midst of a career change. After stints as a student at the Art Institute of Las Vegas and the Academy of Art University in San Francisco, he spent seven years as a freelance illustrator, working on everything from concept art to editorial illustrations. Then came a "straight" job as technical Illustrator for a company that manufactures aircraft and spacecraft. For some time, though, he's been building up his portfolio of scientific illustrations, attending anthropology and paleontology conferences, widening his contacts with working scientists, and drawing for such institutions as the Smithsonian Museum of Natural History and the San Diego Natural History Museum. That's where his true passion lies, and that's the direction he's taking his career in. Whether they're reproductions of museum specimens or reconstructions of what living animals may have looked like, his meticulous paleo-illustrations will convince you, as Catanzaro says, that "science and art, in many ways, really are the same."

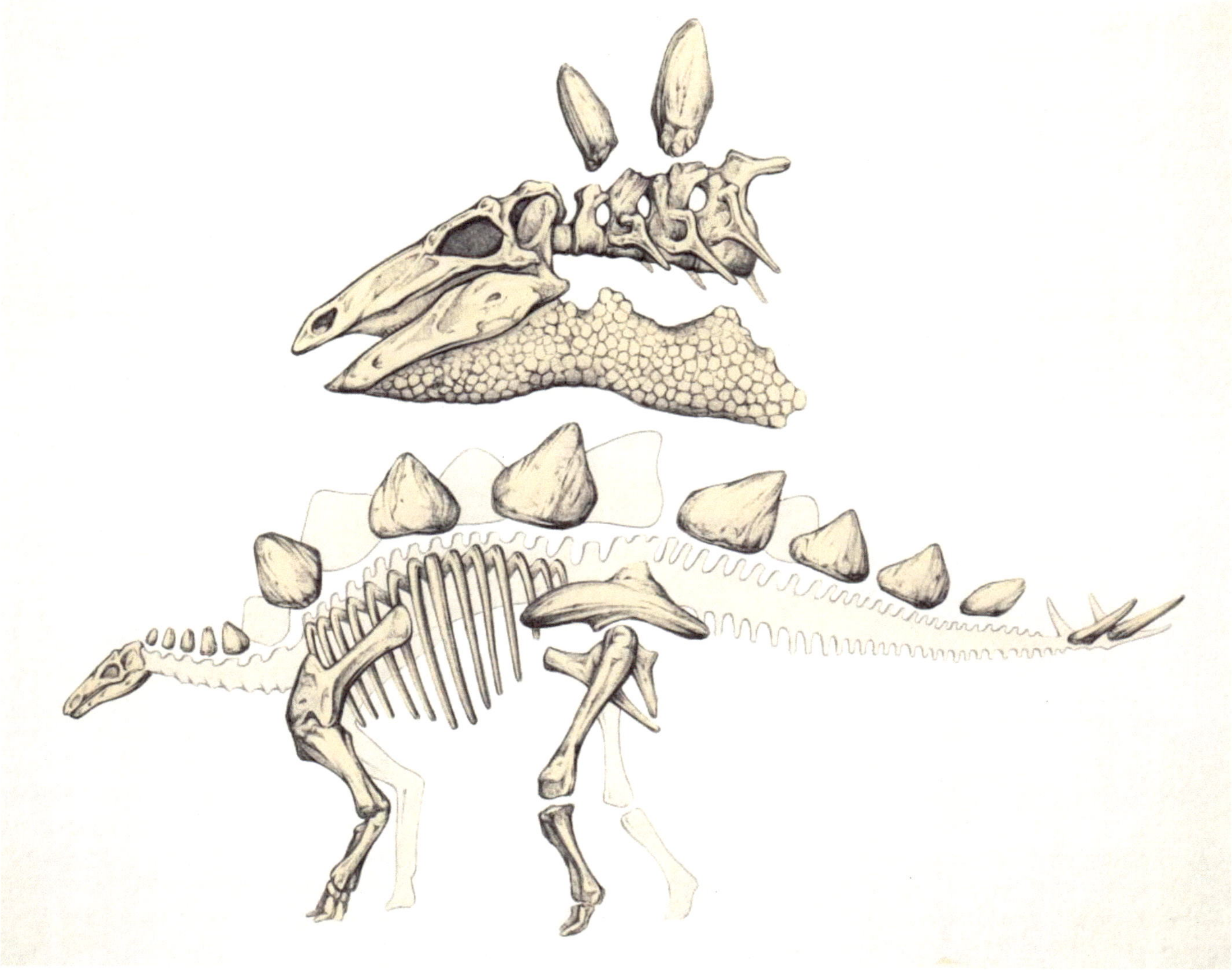

Facing page: *Digital painting of a megalodon (*Carcharocles megalodon*). (Photoshop.)* Above: *Fossil and osteology study of* Stegosaurus. *(Ballpoint ink and Photoshop.) © Jimi Catanzaro; used by permission.*

Fossil News: You're just back from the North American Paleontological Convention (NAPC), where you were displaying and selling your scientific work. Are paleo conferences a good place for scientific illustrators to make connections and show off what they can do?

Jimi Catanzaro: Conferences in general are an excellent place to display your work. They are great for marketing overall actually. You not only have the huge advantage of face-to-face interactions with all sorts of people who might want your services, but you get the opportunity to show your work in person. Seeing your framed, original works on display will always have a greater effect on potential clients than if they receive sample images via email or visit your website.

All of the knowledge and research to which you have direct access at a conference is extremely beneficial, too, and those experiences (the presentations you see and the feedback you receive) ultimately make you a better artist. Possibly the main benefit of attending things like conferences is that it shows dedication to your craft. Lots of people can draw and can even market themselves well online, but if you're an artist who took the time to travel, pay the registration fees, and present your work in person because you have a genuine interest in the subject matter, you're setting yourself apart from the others.

FN: You've mentioned that your hope is to move full-time into professional scientific illustration in paleontology and biology. What prompts that desire?

JC: I suppose my main motivation is to get more involved with the scientific community, and what better way to do that than to use what I love to do most—drawing. This career shift is something I've always wanted to do. Science and art, in many ways, really are the same. For the most part, they both seek an understanding of something and seek a way to represent that information in a way that will educate or inspire others. As an artist, I couldn't think of a better subject

than the natural sciences (biology, paleontology, anatomy). The source material, as well as the inspiration it provides, is almost literally endless. This planet has so much to offer, and there is so much yet to be learned about its history. Being able to make a very small contribution to some of those discoveries one day, and being able to communicate them through illustrations, is truly an awesome way to make a living.

In the meantime, learning, studying, and putting in the work is honestly a lot of the fun! I've spent the last two years building a "scientific Illustration" portfolio, creating new work that represents my passion and displays my competence as a biology/paleontology illustrator. Aside from creating the work, the real goal is getting it out there for the right people to see. Self-marketing is extremely important for an artist—as well as a great deal of work. I send out a lot of emails to museum representatives, magazine editors, authors, paleontologists, biologists, grad students—really anyone who might be in need of an illustrator. Social media has been very helpful, too, and sometimes you can get a lot of attention that way.

FN: What's the relationship between professional paleontologists and illustrators like you? I know, for example, that Julius Csotonyi's work is featured at the Smithsonian in its new Deep Time exhibit—but your work has appeared there as well, hasn't it?

JC: The relationship between paleontologists (or any scientists) and illustrators or artists can vary depending upon the project, but ultimately it's much the same as the relationship between an illustrator of any sort and a client. Drawings are produced with the intention of communicating information of some sort; it's the paleontologist who ensures that the information is as accurate as possible. It's important that illustrators and their clients work together to ensure the artwork maintains a balance between being accurate and informative on the one hand and, on the other, being visually appealing. The artist's goal is not only to help explain and communicate the information or research, but to grab the reader's or viewer's attention. Seeing imagery that aids in describing what a paleontologist might be discussing in an article or publication can really help the reader retain the information.

Meanwhile, I'm flattered to be even loosely included in the same category as Julius Csotonyi! He's an incredible illustrator; I can only hope to achieve that level of notoriety one day. My contribution was much, much smaller, but it was still a very fun project. I was commissioned by a biological anthropologist with the Smithsonian Museum of Natural History to create some osteology and comparative anatomy drawings that were used to illustrate displays and publications at a conference where she was speaking.

FN: I was intrigued when you told me you visited a lot of museums to draw work for your own portfolio. There's something so romantic about that image—the artist with a sketchbook in the museum. At the same time, it speaks to the importance of access to displays of good scientific specimens. Are you able to get "backstage" access?

JC: Unfortunately, I haven't yet gotten "backstage" in any of my favorite museums, but I'm still very fortunate to be able to visit museums to draw and gather photo reference for future drawings [from publicly displayed specimens]. I couldn't agree more that access to these specimens is extremely important. For anyone really, but especially artists. Sure, photos and online resources are great for artists, but you appreciate the specimen so much more when you're standing in front of it, and that definitely contributes to the overall quality of your work. If you don't have access to these exhibits, you miss out on the emotional aspect.

A few years ago, I was at The Natural History Museum of Los Angeles County, gathering reference photos for future projects, when something happened that I had never experienced before. I found myself in front of one of the *Triceratops* skeletons they had assembled and placed on display. As I looked at it and studied it, I was hit with a strange rush of emotions. My hands even began to shake a little! This massive animal .... it was so old and so incredibly beautiful. In its time it was perfectly adapted to its environment, like all of the species we see on display in museums, and it reflected a time when the earth was truly wild. It's something we would know nothing about had it not died under perfect conditions, leaving it preserved in this way. That moment was when I realized just how lucky we all are to have this little window into the past, and I think of that moment pretty much whenever I see a fossil. That to me is the real importance of museums, and it's something everyone should enjoy and appreciate, not just nerdy artists like me.

FN: More than a few of the artists who've allowed us to feature their work have worked with museums. Is that the main career goal, or are there others?

JC: Yes, having my work featured in museums is definitely one of my main career goals! I honestly get excited just thinking about it. Ideally, I would like to do illustration work for museums as well as publications, so I try to market myself for both avenues. I've found that one of the best ways to get the attention of museums is to donate to them. This is something I'm still working on—sort of a side project, I suppose. But one of my favorite things to do, both as a marketing strategy and a way to add to my portfolio when I am not tied up with commissions, is to create illustrations specifically for a particular museum, then mail them the drawing as a donation. It's fun and rewarding for a lot of reasons, but mainly I enjoy it because it's a way to "give back" and show appreciation to some of my favorite museums while still doing what I do best and what I love most. I've done this for a few places, and I plan to do it for many more.

FN: For FN, we're concentrating on your paleo work, but you also illustrate in marine biology, anthropology, and, I'm assuming, other areas. What's been your favorite animal to illustrate so far—even if it's not in paleontology!

JC: That's sort of tough to answer. A favorite animal to draw is almost like asking what's your favorite food to eat—it depends what you're in the mood for! Overall though, if I had to pick just one, I'd have to say cephalopods, mainly octopods and squids. Their anatomy is so unique, which means the approach to drawing them is naturally unique also. All of the basic fundamentals of anatomy drawing sort of go right out the window, mainly because they don't have a skeleton, so you can't use that as a way to find your landmarks as you draw, and they don't have muscle tissue layering over a skeleton the way reptiles, birds, or mammals do. It requires a slightly different approach that can be a lot of fun, and you have a little more freedom to create the drawing however you see fit.

They also make for great subject matter. They have this sort of creepiness to them and, as a fan of nautical monsters and horror literature, I love that. They are one of the few animals that can work just as well for an "infographic style" scientific illustration as they can to illustrate a nautical tale depicting some terrifying creature that sinks ships and pulls unfortunate

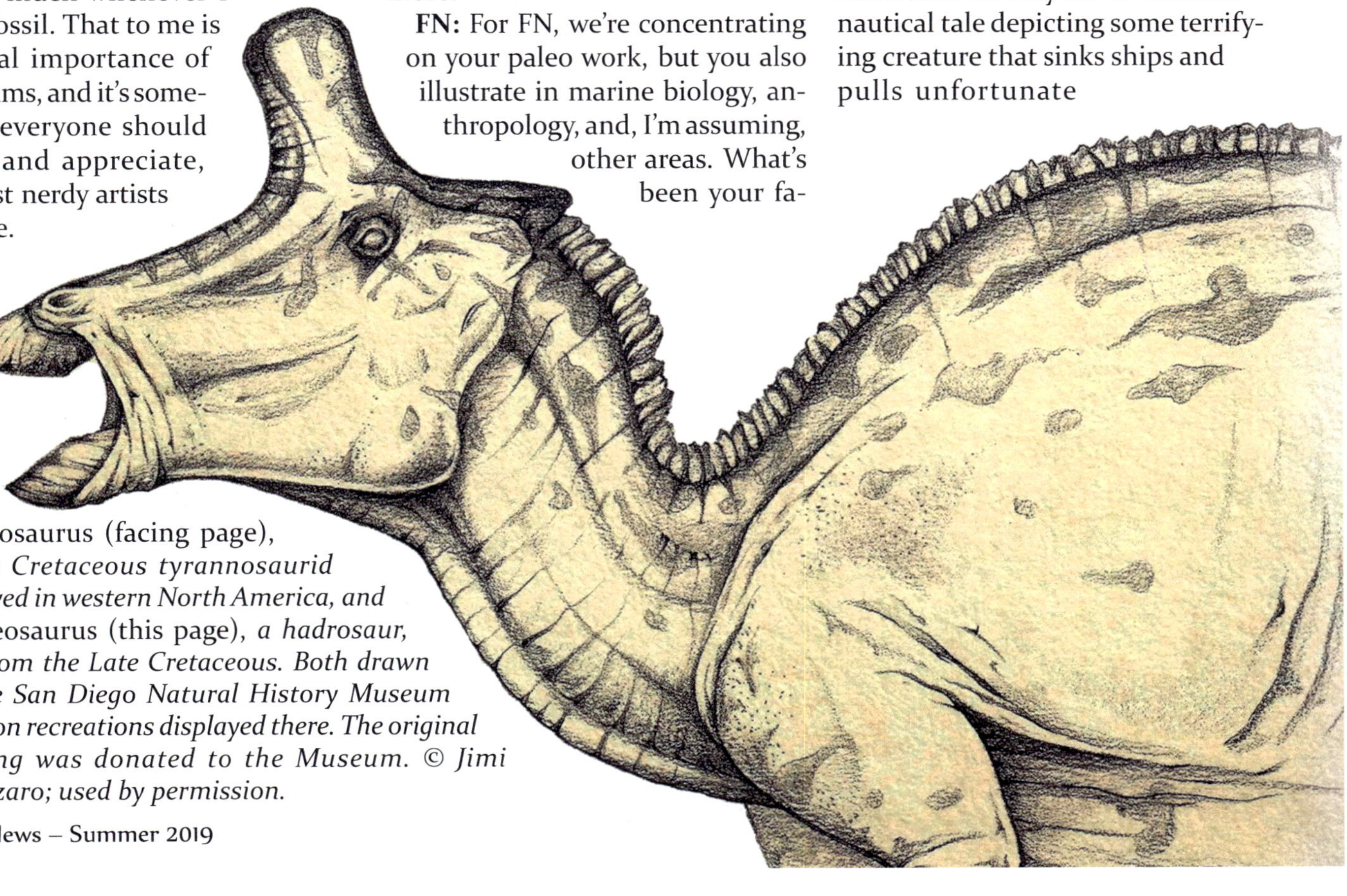

Albertosaurus (facing page), *a Late Cretaceous tyrannosaurid that lived in western North America, and* Lambeosaurus (this page), *a hadrosaur, also from the Late Cretaceous. Both drawn for the San Diego Natural History Museum based on recreations displayed there. The original drawing was donated to the Museum. © Jimi Catanzaro; used by permission.*

sailors to their death. Either way, you can have a lot of fun with them.

FN: Talk about the difference between "scientific illustration" and, in the case of paleontology, "reconstruction" or "restoration" of animals whose appearance we can't really know.

JC: Other people might define it differently but, to me, "scientific illustration" encompasses anything relating to the natural sciences (anatomy, botany, paleontology, anything like that) so really any artwork that would coincide with any scientific publication or any drawing or study that is related to anatomy or biology. This is the side I mostly work in. On the

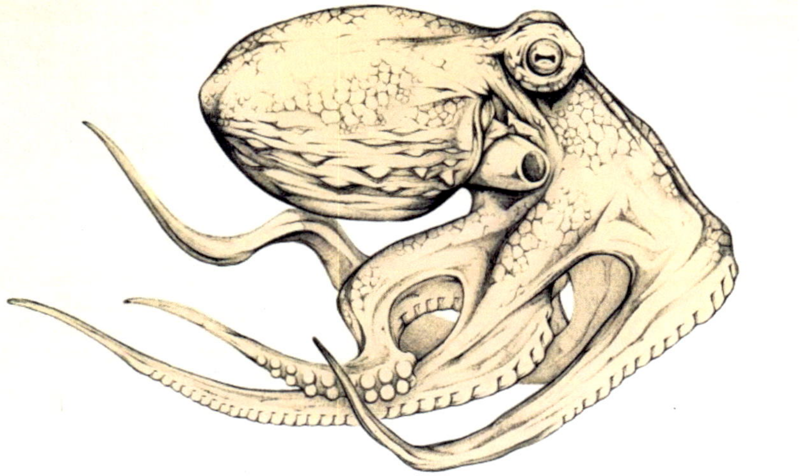

Facing page: *"History of Earth." Illustration depicting some of the animals that have inhabited this planet, both in the distant past and today. (Ballpoint ink on toned paper.)* Above: *A drawing of one of the artist's favorite animals.* Below: *Anatomy and osteology of the coelacanth* Latimeria, *whose ancient relatives belong to the oldest known living lineage of Sarcopterygii (lobe-finned fish and tetrapods), making them more closely related to reptiles and mammals than to common ray-finned fish. (Ballpoint ink and Photoshop.) © Jimi Catanzaro; used by permission.*

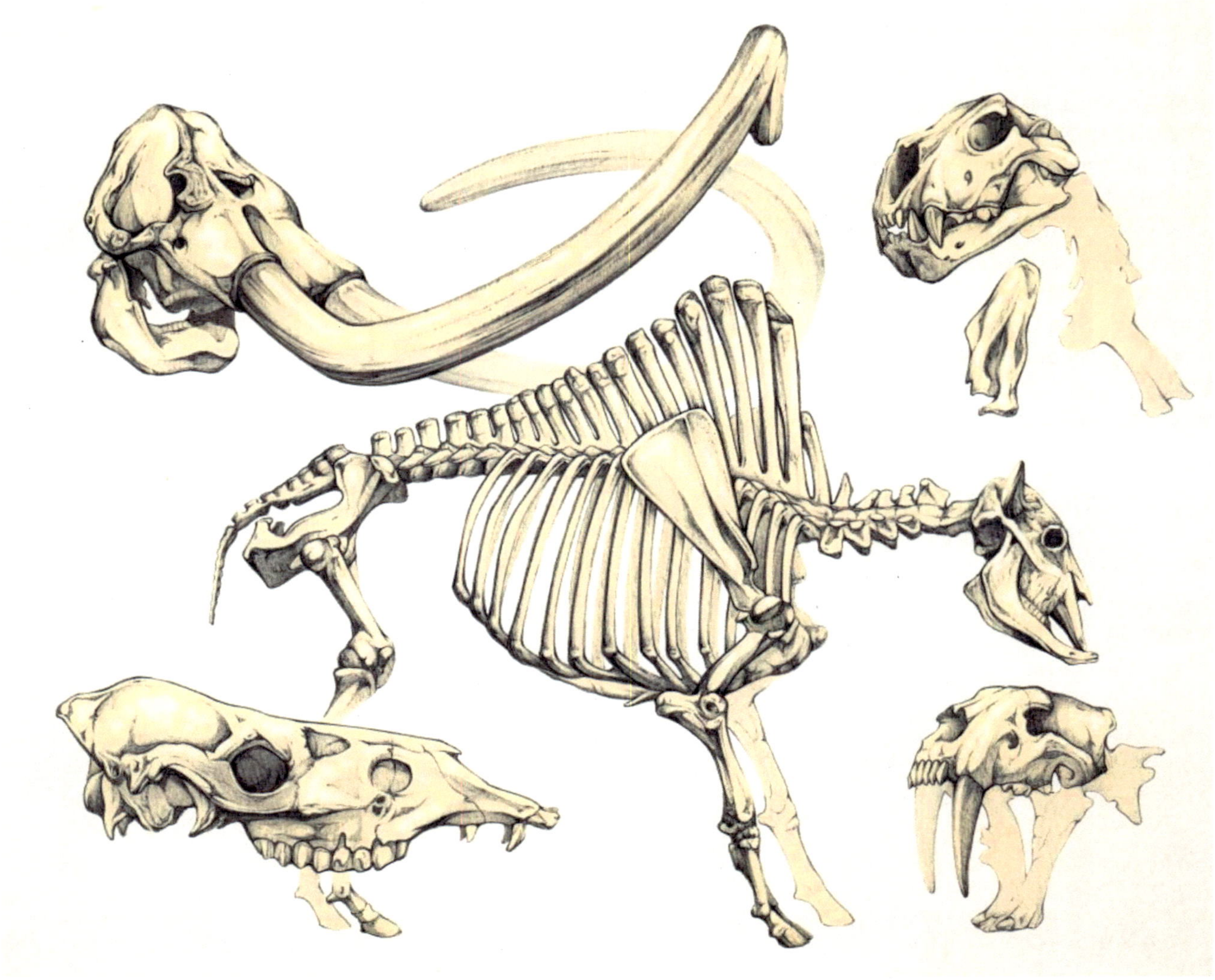

This is a drawing of some of the fossilized mammals that have been discovered in the La Brea Tar Pits and which are on display at the La Brea Tar Pits Museum in Los Angeles, California. This piece serves as a good example of the way I prefer to work. While on a visit to the museum, I took advantage of the opportunity to do my absolute favorite thing aside from drawing: museum exploring! When visiting museums, I always make it a point to study each exhibit and display closely and to retain as much information as possible. I truly feel that doing this has made me a better artist and has helped me develop a solid foundation and a good understanding of anatomy, imperative for producing this sort of artwork. I also gather as many reference photos as possible—usually hundreds of them! Sometimes these photos get put to use right away for upcoming projects; other times they come in handy years later. Either way they're good to have and, as an illustrator, I find it's always best to be able to rely on your own photography for reference. Before I had even left the museum, I was already inspired to create my next set of fossil studies. Back home, I looked through the photos and decided which of the mammal species were most aesthetically pleasing and best represented the array of specimens at the museum. After many sketches, rough drafts, and studies, I created this drawing of what I considered the museum's most inspiring and beautiful fossils.

— J.C.

other hand, "recreations" or "restorations" are far more complex illustrations, in my opinion, and require a higher degree of knowledge. I have much less experience working in recreation-style paleoart, and I still have so much to learn. There are some truly brilliant artists who have mastered this particular field. I think of James Gurney, John Gurche, and John Sibbick. They are so highly skilled that they really belong in their own category. To create illustrations effectively in that style, you not only have to use the tools, theories, and concepts that all artists do (composition, color theory, anatomy, and so on) but you also have the added obstacle of finding the perfect balance of established scientific theories and imagination to provide that small glimpse into the past through your art. It's the closest thing we have to "going back in time," and I've always been obsessed with that idea.

Above: *Catanzaro at his desk, working from preliminary studies and consulting a copy of* Prehistoric Life: The Rise of the Vertebrates *by David Norman with illustrations by John Sibbick. Photo © Ashley Catanzaro; used by permission.*

David CORDIE

Recent Papers in Paleontology

Humans have long been attracted to the soft golden glow of amber. Since prehistoric times, societies have used amber to adorn everything from necklaces to caskets. Even movies have highlighted it—in *Jurassic Park,* amber served as the vehicle for bringing dinosaurs back to life. The movie explains that "dino DNA," trapped in the abdomens of mosquitoes contained within amber, was used to synthesize the park's creatures. The character John Hammond even carries a piece of amber atop his cane.

The Cretaceous Burmese amber specimen with its numerous inclusions—and the remarkable ammonite. Figure taken from Yu et al. (2019).

The movie may have exaggerated the properties of amber a bit. First, amber is very soft (a hardness of ~2 on the Moh Scale of Mineral Hardness) and could easily be scratched and chipped. Definitely not the best material for a walking staff. Second, the film describes amber as coming from hardened tree sap. More accurately, it comes from tree *resin.* Sap is a fluid used by plants to transport nutrients within their internal xylem (the vascular system of a plant). It's also collected by humans for making syrup. Resin, on the other hand, is a highly viscous material found in the external, woody portions of the plant and is used as a defense mechanism against parasites and insects. Finally, DNA just isn't stable enough to last for sixty-five million years, even in amber; even if DNA were retrievable in that situation, it would most likely belong to the mosquito rather than to a dinosaur.

This doesn't mean that amber can't produce some amazing finds, however. Amber, as described above, first starts as resin, which is a mix of organic resinous material and naturally occurring alcohols. The sticky organic material wards off boring insects (due to the risk of entrapment), but also flows as a fluid because of the liquid alcohol. Gradually, the alcohol evaporates, and the remaining sticky substance begins to solidify into copal. The gummy copal is often clearer than amber (ancient peoples used it as incense). As this substance continues to harden, polymerization (the tangling of large organic molecules) occurs, and relatively harder amber is left.

Complex sutures like the ones in this fragment of an ammonite shell resemble a Christmas tree. They helped strengthen the chambers of the shell so the organism could change levels in the ocean through a gas-exchange system—diving into deeper water, rising closer to the surface, or simply hovering at a specific depth. Their arrangement and shape is used to identify ammonite species. Living Nautilus *also have sutures, albeit very simple ones, but they use the same system to move horizontally in the ocean and change buoyancy, filling the chambers of their shells with gas and then emptying them.*

Yellow amber is most common, but red and even extremely rare blue ambers from the Dominican Republic have been found. Most amber comes from coniferous trees and, because of the long hardening process, date to the Cretaceous or Paleogene (between 140 to 30 million years ago). Of interest to paleontologists is the fact that the original resin often captured tree-dwelling insects as they crawled along the tree trunks. Other, more unique finds have also occurred, though—entire lizards trapped in amber or the tails of feathered dinosaurs (Sherratt et al., 2015; Xing et al., 2016).

Such organisms as these were, at least, terrestrial in nature. A new paper published in mid-2019 in *Proceedings of the National Academy of Sciences* (*PNAS*) reported that even marine organisms could have been caught in these sticky traps.

In northern Myanmar, a small, six-gram piece of yellow amber was found. Dating of surrounding zircons suggested it was from the Cretaceous. The specimen contained a number of arthropod specimens typical of amber from this region, including twenty-three specimens of mites, a spider, millipedes, cockroaches, beetles, flies, and wasps. At least forty individual arthropod specimens are present, all of them forest-floor-dwelling organisms generally associated with leaf litter. Given the terrestrial nature of most trees and the natural pull of gravity on fluid resin, it is not surprising to find these types of organisms in amber. In addition, at least four specimens of isopods were recognized in the amber specimen. Some of them had reduced eyes, a common characteristic for terrestrial isopods. The preservation of these specimens

Yu, T., Kelly, R., Mu, L., Ross, A., Kennedy, J., Broly, P., Xia, F., Zhang, H., Wang, B., Dilcher, D. (2019, June). An Ammonite Trapped in Burmese Amber. *Proceedings of the National Academy of Sciences 116*(23): 11345–11350. Available at https://doi.org/10.1073/pnas.1821292116.

was not complete, however, and the differences noted in their eyes might have been a taphonomic effect. Regardless, the body morphology of other isopod specimens within the amber made them clearly identifiable as marine or intertidal in origin. Given the well-defined terrestrial arthropods mentioned above, it seems likely that the tree that produced this resin was in a coastal forest.

Here is where the amber assemblage gets interesting. Four marine gastropods and a small ammonite were also present within the amber. The gastropods were well enough preserved to be identified as belonging to the genus *Mathilda*. This genus has a recognizable conical shape with tight whorls and is only found in marine environments. The ammonite measures 33 mm long, 9.5 mm wide, and 29 mm high, and its distinctive sutures are visible in microtomography. The subclass Ammonoidea is an extinct subclass of cephalopods that includes octopus, squid, and cuttlefish.

> The authors propose three possible scenarios to explain how a marine organism could end up amongst the trees.

A closeup of the curious ammonite in Burmese amber. Figure taken from Yu et al. (2019).

Nearly all modern representatives of cephalopods have lost their external shells (save for *Nautilus*) but, in the past, their chambered shells housed the more typical bundle of tentacles we associate with cephalopods today. Chambers within the ammonite's shell were separated by sutures, and the shape and complexity of the suture lines are useful for identification. The most complex suture types, termed ammonitic sutures, look like groves of Christmas trees. Thus, the clearly visible ammonitic sutures within the Burmese amber instantly identify this organism as a marine ammonite. The authors note that this is the first time a macroscopic marine shell has ever been found in amber and offers interesting clues as to the location of the forest that produced it.

The authors, Tinting Yu and others, propose three possible scenarios to explain how a marine organism could end up amongst the trees. First, a terrestrial tree overhanging a sandy beach could have produced resin that slid down into a pool of marine water and just happened to trap a passing ammonite. Second, a large tsunami could have flooded a forest, allowing an ammonite to venture inland and get captured by resin. Finally, a dead ammonite shell could have washed ashore and been blown inland during a tropical storm. Given that the amber and the aperture of the ammonite contained coarse sand grains, the tree and marine environment must have been close together and, thus, the authors currently favor the first scenario. The fact that liquid resin solidifies poorly when submerged in water could answer why this type of preservation for marine organisms is not more common in the fossil record, despite the regularity of coastal forests.

We can hope that future finds of marine fossils in amber will shed more light on the dynamics of coastal forests and amber preservation. For now, though, we should be content that the fossil record does occasionally provide rare glimpses into the past that no one ever expected to find.

No mosquitoes required.

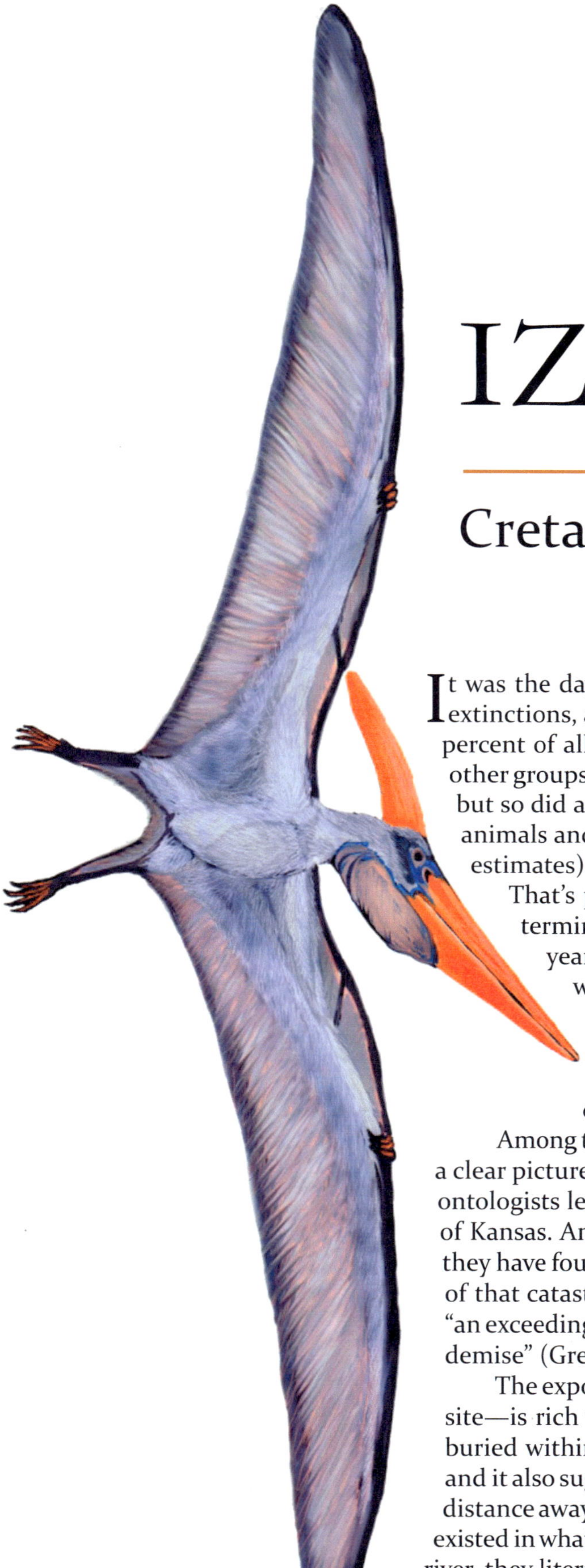

Yasmani Ceballos IZQUIERDO

Cretaceous Pterosaurs in Cuba?

It was the day that marked the beginning of one of the planet's great extinctions, a day that heralded the end for something like ninety-nine percent of all living creatures on Earth. Most famously, dinosaurs and other groups of giant reptiles disappeared in the subsequent extinction, but so did ammonites and countless other species of land and marine animals and plants (as much as eighty percent of all species, by some estimates).

That's perhaps the main reason why scientists have been so determined to understand exactly what happened sixty-six million years ago, when a huge asteroid plunged into a shallow sea near what is today Mexico. The impact bombarded the planet with scalding rock and molten glass, sent literal tons of dust into the atmosphere, and precipitated a global disaster that included acid rain and wildfires that are estimated to have consumed some seventy percent of the planet's forests.

Among those who imagined a time when it would be possible to paint a clear picture of the infamous K-Pg extinction "event" is a team of paleontologists led by Robert DePalma, a doctoral student at the University of Kansas. And now their dreams seem to have come true. They believe they have found, in North Dakota, the remains of the immediate victims of that catastrophe in what appears to be a genuine open-air museum, "an exceedingly rare snapshot of the moment that marked the dinosaurs' demise" (Greshko, 2019).

The exposure—christened "Tanis" by DePalma, who discovered the site—is rich with the fossils of animals and plants that died and were buried within as little as an hour following the impact of the asteroid, and it also suggests the consequences the impact had on areas even some distance away. The shock generated giant waves in an inland sea that once existed in what is today North Dakota. When the waves reached an adjacent river, they literally reversed its direction. Torrents of stones and debris fell upon the fish in these waters and upon plants, insects, and other land animals, all of which were then buried in the sediments carried in subsequent waves. Over time, they fossilized in an identifiable layer.

Geophysicist Mark Richards believes that the "worldwide earthquake generated by the KT impact could have been a thousand times stronger than the biggest earthquake ever experienced in human history." Though scientists say that the actual tsunami would have reached Tanis seventeen hours later, Richards has "calculated that potent seismic waves would have arrived at [the site] six minutes, ten minutes, and thirteen minutes after the impact" (Preston, 2019). Those seismic waves would have carried the energy necessary to create a powerful back-and-forth sloshing, like water in a bathtub, called seiche waves. At Tanis, the seiche waves are estimated to have grown as tall as ten meters (thirty-three feet). DePalma and other paleontologists believe they are responsible for the remarkable fossil deposits there.

In a *New Yorker* article on the Tanis find, Douglas Preston imagined those early moments this way:

> the water ... deposited everything that had been caught up in its travels—the heaviest material first, up to whatever was floating on the surface. All of it was quickly entombed and preserved in the muck: dying and dead creatures, both marine and freshwater; plants, seeds, tree trunks, roots, cones, pine needles, flowers, and pollen; shells, bones, teeth, and eggs; tektites, shocked minerals, tiny diamonds, iridium-laden dust, ash, charcoal, and amber-smeared wood (2019).

DePalma, in fact, reported that the North Dakota site even yielded flooded ant and mammal burrows and "the thigh bone of a large sea turtle" (Preston, 2019). Questions about some of these claims, however, began to be raised after DePalma's official publication of the first analyses of the Tanis site (see, e.g., Mitchell, 2019).

Previous page: Pteranodon. © *Dmitry Bogdanov; used by permission. This page, top: Mass of articulated fish from the Tanis inundation surge deposit.* Below: *Magnified images of the fish above, enhanced to show minute tektites clustered within the gill region. Photos courtesy of Robert DePalma/University of Kansas; taken from DePalma et al. (2019).*

Fish had time to breathe in tiny fragments of hot glass—propelled through the air by the asteroid impact and then dropped into the water. These minute bits of glass, known as tektites, were later found fossilized in their gills.

At the University of Amsterdam, meanwhile, Jan Smit announced that he and his colleagues had discovered the tracks of dinosaur left just moments before the animals' death in the post-asteroid catastrophe. According to Smit, traces of a carnivore, possibly a juvenile Tyrannosaurus rex, are "definite proof that the dinosaurs were alive and kicking at the time of impact.... They were running around, chasing each other" when they were suddenly overcome (Borenstein, 2019).

Prehistoric Rodas

Even as the team of scientists in North Dakota worked to decipher evidence of the "perfect catastrophe"—and their efforts continued to make headlines—news of another important fossil discovery reached the international scientific community. This time, the news came from Cuba and, specifically, from the Damují Paleontological Region, located near Rodas (in the Province of Cienfuegos in east-central Cuba).

Above: *One of the specimens* in situ *in the arroyo.* Below: *A closer look. Photos © Carlos Rafael Borges Sellén; used by permission.*

The Cuban fossils—two bones, one measuring 3 cm x 1.5 cm and the other 4 cm x 2.5 cm (approximately 1.1" x 0.6" and 1.5"x 0.98")—were discovered by Researcher Carlos Rafael Borges Sellén in an area south of Rodas, ground that he and researcher Alberto Arano Ruiz had studied for years, logging a number of significant contributions to the fossil record of the Cretaceous in Cuba and the Caribbean generally.

Lázaro William Viñola, a student in the Department of Earth Sciences at Montana State University, and his colleagues were the first to suggest that the specimens might belong to a flying reptile or pterosaur, yet another of the doomed creatures that disappeared in the end-Cretaceous extinction. With the help of Dr. Manuel Iturralde-Vinent, Viñola and his group contacted Dr. Zulma Gasparini of the Museo de la Plata in Argentina and Dr. Laura Codorniú, a fossil vertebrate specialist also from Argentina, who confirmed that the fossils had, indeed, come from a pterosaur.

Pterosaurs were flying reptiles that lived from the end of the Triassic, some 205 million years ago, until they went extinct at the end of the Cretaceous. Although it's common to hear pterosaurs referred to as "flying dinosaurs," in fact they weren't dinosaurs at all. Current

"Life restoration of a group of giant azhdarchids, Quetzalcoatlus northropi, *foraging on a Cretaceous fern prairie. One pterosaur has procured a juvenile titanosaur, while the others stalk through the scrub in search of small vertebrates and other foodstuffs." Image © Mark Witton; used by permission. Azhdarchids are a family of large, Late Cretaceous pterosaurs. Scientists don't yet know whether the Rodas bones are from an azhdarchid, but azhdarchids were among the last known surviving pterosaurs. Pterosaurs disappeared entirely from the fossil record in the end-Cretaceous mass extinction.*

scientific opinion, however, is that pterosaurs are more closely related to birds and dinosaurs than to any living reptile.

Pterosaur expert Fabio Dalla Vecchia, of the Catalán Paleontological Institute in Spain, described one of the Rodas specimens, noting that "the fragment may be from a large limb bone." He continues, "It is elliptical in cross-section and has an extremely thin cortex (which indicates that it is a pterosaur bone). The hollow in the interior of the bone was filled with sand that, through a process of cementation, became sandstone. The cortex is not completely preserved over the entire surface of the fragment, which suggests that the specimen was reworked.

"The identification of this fossil bone has been quite difficult. Obviously, it comes from a pterodactyloid pterosaur of moderate size, though it would only be speculation to try to establish the animal's [exact] dimensions. By studying a thin section of the cortex, [we could] establish whether the bone belonged to an adult or to a juvenile," Dalla Vecchia adds.

The Cantabria Connection

Though they don't represent the exact moment in which dinosaurs died, the Rodas fossils, which are notable for their age, offer clues to a marine environment between seventy to sixty-six million years ago (at the end of the Cretaceous). According to geologists, the Cuban fossils belong to the Cantabria Formation—the same zone that, in 2018, yielded the four-centimeter fossil of a starfish, the oldest know from Cuba, and which had already produced a crustacean of the Family Raninidae (known commonly as "frog crabs") as well as bivalves and small rudists.

The pterosaur bones were found near an arroyo in which the Monos Formation is exposed and, above it,

Above: *Close-ups of the Rodas specimens; scale in millimeters.* At right: *A transverse view. If the bone were thin-sectioned, details of the cortex could be studied (notice the thin bone layer, especially visible on the right edge of the specimen in this image), which would provide a more definitive identification of the fossils. Photos © Carlos Rafael Borges Sellén; used by permission.*

the Cantabria Formation. During heavy rains, the sediments are reworked and a great deal of tumbled material comes to the surface. According to Ruiz, sediments of terrestrial origin (i.e., an island) can also be inferred from the site because of the presence of a number of fossil tree trunks. At the time, the island was evidently subjected to a marine transgression. As a result, many of those tree trunks were invaded by wood-eating organisms (such as *Teredolites*, an ichnogenus believed to be associated with bivalve borings), and that has helped identify a marine layer above the terrestrial sediments.

The Damují Paleontological Region has also yielded the tooth of a mosasaur (a carnivorous marine reptile), discovered by Borges Sellén in June 2015. According to Michael J. Everhart, a mosasaur expert at the Sternberg Museum of Natural History in Wichita, Kansas and the author of *Oceans of Kansas*, the tooth constitutes "a major find at the end of the Cretaceous. Indeed, it might be from one of the very last mosasaurs...."

Dalla Vecchia notes that the "importance [of the Rodas site] is predominantly paleogeographical because it extends the distribution of large Cretaceous pterodactyloids to Cuba." Indeed, the area has yielded the only evidence so far of the last saurians in this part of the Caribbean. Perhaps the most interesting thing the Rodas site offers, then, is a glimpse of the prelude of extinction.

[English translation and additional material & research by Wendell Ricketts.]

Recommendations for Further Reading & Works Cited

Borenstein, Seth (2019). Fossils Show the Day a Killer Asteroid Wiped out the Dinosaurs, Researchers Say. *Chicago Tribune*. https://www.chicagotribune.com/news/environment/ct-asteroid-dinosars-extinction-20190329-story.html.

DePalma, Robert A., Smit, Jan, Burnham, David A., Kuiper, Klaudia, Manning, Phillip L., Oleinik, Anton, Larson, Peter, Maurrasse, Florentin J., Vellekoop, Johan, Richards, Mark A., Gurche, Loren, and Alvarez, Walter (2019, April). A Seismically Induced Onshore Surge Deposit at the K/Pg Boundary, North Dakota. *Proceedings of the National Academy of Sciences, 116*(17): 8190-8199. DOI: 10.1073/pnas.1817407116.

Greshko, Michael (2019, 31 March). These Fossils May Capture the Day the Dinosaurs Died. *NationalGeographic.com*. https://www.nationalgeographic.com/science/2019/03/fossils-found-from-day-dinosaurs-died-chicxulub-tanis-cretaceous-extinction.

Mitchell, Alana (2019, 9 May). Why This Stunning Dinosaur Fossil Discovery Has Scientists Stomping Mad. *MacLeans.ca*. https://www.macleans.ca/society/science/why-this-stunning-dinosaur-fossil-discovery-has-scientists-stomping-mad.

Preston, Doug (2019, 29 March). The Day the Dinosaurs Died. *New Yorker*. https://www.newyorker.com/magazine/2019/04/08/the-day-the-dinosaurs-died.

WYOMING FOSSILS
Sawaya's Center
921 Pine Avenue
Kemmerer, Wyoming
(254)223-3204
visit our website www.wyomingfossils.com
OPEN DAILY
9:00 a.m.
to 6:00 p.m.
• Fossils
• Minerals
• Crystals
• Unique Items
• Gifts
❖FIND US ON FACEBOOK/INSTAGRAM UNDER WYOMING FOSSILS

A Voyage Through Italy's Ancient Tropical Forests

Nando MUSSMARA

The Hall of Palms

As a young boy in Italy, I was obsessed with Giotto crayons. They were the most popular brand—like Crayola in the U.S. —and anyone who was a child in Italy at the time almost certainly had them. In particular, I was fascinated by the design on the box: a colorful illustration of a young Giotto, the master Italian painter of the Late Middle Ages, drawing a lamb on a slab of rock while his master, Cimabue, looks on admiringly. ¶ At the beginning of every school year, I tormented myself with the same questions: Do I need the box of six colors or the one with twelve? Will I really use all those, or could I make do with the small box plus the stubs of crayons left over from the previous year? Is a green crayon indispensable, given that I can make the same color by mixing blue and yellow? ¶ By the time I got to high school, my studies in the classics and the humanities ensured that I would never forget Giotto di Bondone, the man who ushered in a genuine revolution in the history of painting. He was, after all, the artist who introduced the properties of space and perspective to painting, making it possible for artists to capture vivid, lifelike, three-dimensional subjects. Giotto paid special attention to the details of facial expressions, and he meticulously observed bodies in motion in order to locate them precisely in space. To put it more simply: In a single blow, he put an end to the static, immobile subjects painted by the artists who came before him.

One of the greatest of Giotto's masterpieces is housed today in the Scrovegni Chapel in Padua, a mid-sized industrial town in Northeast Italy. Constructed between 1300-1305, the chapel was frescoed entirely by Giotto. Following a major earthquake that hit the area in 1976, however, the chapel became unstable and was closed to the public—for nearly a quarter of a century—while major repair and restauration work went on. Meanwhile, I'd continued to dream about seeing Giotto's frescoes in person.

These are the memories that came back to me recently when I happened to come upon a shepherd who, to help time pass, was sketching his flock as it grazed in a grassy field. I also noticed two palms in the field, survivors of the widespread destruction caused in Italy by an infestation of the red palm weevil (*Rhynchophorus ferrugineus*). What a pity to see these magnificent trees, whose fossil record goes back tens of millions of years, succumb to a tiny red insect (the weevil is actually orange and not red, as long as we're talking about colors).

The shepherd, the drawing, Giotto, palms ... Knowing that the city of Padua houses not only the Scrovegni Chapel but a world-class paleontological museum with an entire hall dedicated to fossil palms, I began to hatch the idea of an expedition: between Giotto and Middle Eocene fossil palms, I could kill, as the rather violent saying goes, two birds with one stone.

The Sala delle Palme

The Sala delle Palme ("Hall of Palms"), part of the University of Padua's Museum of Geology and Paleon-

Facing page: *The illustration from the seventies-era Giotto crayons box. The young Giotto draws a lamb as his mentor, Cimabue, looks on.* Above: *The first display case that greets visitors at the entrance to the Hall of Palms. Three examples of palm fossils complete with trunks and more than two meters (about 6-1/2 feet) high. Purga di Bolca site, Middle Eocene. Unless otherwise stated, this and all photos in this article © Nando Musmarra; used by permission.*

tology, is housed in Palazzo Cavalli, a sixteenth-century villa. The palazzo and the Hall of Palms were restored not long ago to ensure, among other things, optimal conditions for the conservation of its specimens. Visitors can not only enjoy the beauty of giant palms but wander through a magnificent fossil forest and imagine it as it was millions of years ago. The University of Padua dates to 1222, but its paleobotany department—whose collections are integrated into the displays in the Hall of Palms—houses specimens that are 300 million years old. They come largely from sites in the Triveneto area period) and were all collected from famous exposures near the town of Bolca (Province of Verona) in the Veneto Region and along the Chiavon Valley (Province of Vicenza).

The Evolution of Plants: A Brief History

Though many readers may be familiar with the evolution of plant life, it's worth tracing that history (with special attention to the Subalpine area of Northeast Italy) in the way the museum does in its displays in the Hall of Palms.

Algae, microfossils, and columns and mounds of

View of one of the exhibition rooms in the Hall of Fossil Palms.

(the Triveneto is a historical region of Italy that covers the modern-day regions of Veneto, Friuli Venezia Giulia and Trentino-Alto Adige/Südtirol; most commonly, it is simply called "Northeast Italy"). In general, the paleobotany collection documents the evolutional history of plant life through specimens gathered from important sites all over the world.

The palm fossils, meanwhile, are roughly thirty to fifty million years old (dating to the Eocene-Oligocene

stromatolites lived during the Precambrian, Cambrian, and Ordovician, sometimes literally creating the depositional surface for later fossils. It wasn't until the Silurian, however, that the newly exposed dry land was colonized by plants, including lichens, mushrooms, and mosses.

The first real land plants were small and quite simple, though they were already equipped with minute leaves. They also had a vascular system that allowed fluids to conduct water, dissolved minerals, and food to the leaves

1 YEAR ANNIVERSARY SALE
For Online Store
PaleoJoe: The Fossil Store
Museum and High Quality Fossils
One of a kind fossils
Dinosaur Bones, Dinosaur Teeth, Trilobites, Fish, Crinoids, Ammonites, Ferns, Cystoids, Dinosaur Posters, Shark Teeth, Childrens books ....so much more.
"Thank you for the speedy delivery. They look great along with the other 18 specimens I have." -David, TX
"Thanks. I love my new Trilobite, and that beautiful matrix it is in, helps make it the perfect piece to liven up my boring office cubicle" -Lawrence, Michigan
"I have just got the fossil. It is superb!" -Manuel, Brazil
"I really love your products and business practices" -Janice, West Virginia
Shop PaleoJoe.com The Fossil Store
Enter code: 1year for a 15% OFF Sale to celebrate our 1st year of business.
All orders over $100 include free shipping in the USA
Contact Jen at
or PaleoJoe at
or visit the website PaleoJoe.com for more Fossil Fun

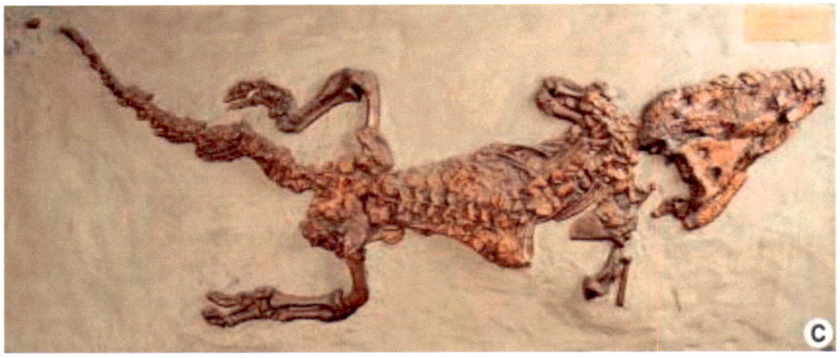

A sampling of the spectacular fossils recovered from the Bolca (Province of Verona, Italy) area. At right: *A) Holotype of* Geonomites saturnia *De Visiani 1864. From the Vegroni site, northeast of Bolca. B)* Latanites *sp., Purga di Bolca site. Both are in the "Hall of Palms."* Above: "Crocodilus vicetinus" *Lioy 1865. Replica of the holotype, collected from Purga di Bolca, that was destroyed during WWII; the replica is now held at the Museum of Geology and Paleontology of the University of Padua. The Bolca sites originally yielded ten crocodilian specimens when the area was first studied in 1865 and then, thirty years later, five more. Freshwater turtles and a snake have also been found. Images taken from Giusberti, Del Favero, and Roghi (2014).*

and other structures of the plant. Just as today, these ancient plants were protected against water loss by a membrane, known as the cuticle (a waxy covering that can be found on almost all exposed surfaces of a plant). Tiny pores in the cuticle (called stomata) inhibit the loss of water—in other words, they make sure the plant doesn't dry out. Sexual production took place via spores.

During the Devonian, an important process of diversification led to the development of trees with actual trunks. In fact, representatives of almost all the plant groups that exist today, excluding flowering plants, evolved then.

The continued explosion of diversity in forms of plant flora is evident in Carboniferous plant fossils. A favorable climate and the presence of marshes and swamps allowed lush forests to emerge, filled with grassy plants, horsetails (*Equisetum*), lycopods, ferns, seed ferns (pteridosperms or, as they are more commonly known, pteridospermatophytes), and Cordaitales (early conifers, now extinct).

With the dramatic changes in climate and in the position of the land masses during the Permian, plants that were typical of wetland environments declined as others, more adapted to drier climates, began to proliferate. Lycophytes went extinct even as conifers began to diversify, and the first primitive cycads and the first ginkgophytes appeared. Early seed-fern species (pteridospermatophytes) were replaced by modern forms better adapted to drier environments.

In the Italian subalpine area, plants of Permian age have been found in a number of localities, such as Egna (Province of Bolzano) and Recoaro (Province of Vicenza) along with a fossil flora that includes simple conifers and the sparse remains of primitive ginkgophytes, seed ferns and, more rarely still, fragments of the fronds of so-called true ferns.

The great end-Permian mass extinctions spared few groups of primitive plants, including, for example, the Cordaitales (early conifers), and the diversity and number of plant species in general dropped dramatically. As always happens after a great catastrophe, a transitional period followed, during which plants had to play catch-up to survive, and then vegetable life once again experienced a renaissance.

During the Middle Triassic, various groups appeared that would give rise to modern plant families. Of particular relevance was the descent of the bennetitalians (a now-extinct group, once considered a predecessor of flowering plants) and the return of the cycads which, along with conifers, literally exploded in number and diversity, ushering in the age of gymnosperms (plants with seeds).

The fossil jewel of the vegetable world of Northeast Italy is amber, found in small droplets in deposits that date to the middle part of the Dürrenstein Formation (Late Triassic, about 230 million years ago). Dolomite amber is among the oldest in the world and has been recovered mainly around the town of Cortina d'Ampezzo in the southern (Dolomitic) Alps of the Veneto Region.

During the period that followed, the Jurassic, no important new groups of plants appeared, but a major differentiation within existing groups began to take place. Jurassic floras were, in general, dominated by cycads, evergreens such as *Araucaria*, and resurgent bennetitalians. In more humid areas, true ferns abounded, as did seed ferns and ginkgophytes.

In Northeast Italy, one of the most famous Jurassic floras comes from the Rotzo Formation of the Calcari Grigi Group of the Venetian Prealps. The so-called "Rotzo Flora," known both from the Vicenza area

as well as from around Verona, was studied and described for the first time by Baron Achille De Zigno between 1856 and 1885 in his celebrated work, *Flora Fossilis Formationis Oolithicae* ("Oolitic Fossil Plants"). His collection of over 3000 specimens of Italian fossil plants is how housed at the Museum of Geology and Paleontology in Padua.

During the Early Cretaceous (the last period of the Mesozoic), the most significant event was without doubt the appearance of angiosperms (seed-bearing, fruit-producing, flowering plants). Though angiosperms are today the most important and most diverse group of plants on the planet, during the Early Cretaceous they were small, grassy varieties that grew in humid and swampy areas. They quickly became the dominant group of plants, and their descendants ultimately included the many fruit-bearing trees we recognize today. At the same time, the fossil record tells us that there was a reduction in the bennetitalians, cycads, and seed ferns which had, up to that point, dominated Mesozoic floras.

The Eocene is particularly well represented in Northeast Italy. The most famous of the Monte Bolca (Province of Verona) exposures are undoubtedly the Pesciara, the Spilecco, and the Monte Postale deposits, which

This and previous page: *Illustrations of species named and described for the first time by Baron Achille De Zigno and published in his* Flora Fossilis Formationis Oolithicae *("Oolitic Fossil Plants"), 1856-1885. Previous page:* Dichopteris visianica *Zigno. A pteridosperm (seed fern) from the Lower Jurassic "Calcari Grigi" of the Venetian Prealps. This page, left: Large frond of* Danaeites heerii *Zigno, a fern or cycad. Below:* Sagenopteris goeppertiana *Zigno, a pteridosperm. Facing page, at right, one of the beautifully preserved fish from the Bolca "Fishbowl" Lagerstätte, Pesciara di Bolca, and on display at the Museum of Geology and Paleontology in Padua. Shown is* Platax subvespertilio, *a batfish. At left is Louis Agassiz's illustration of* P. macropterygius *from his famous* Recherches sur les Poissons Fossiles, *Vol. 4, 1855, Plate 41a.* P. macropterygius *is also known from the Bolca sites, and species in the genus are still alive today in the Western Pacific.*

have yielded spectacular fossils of fish and other fauna (corals, mollusks, crustaceans, insects, bird feathers, and rare traces of snakes) that date back some fifty million years (to the Lower Eocene).

What is striking about these deposits is not only the wealth of species and of specimens they produce, but also the fossils' perfect state of preservation. Most of the Bolca fish fossils have "relatives" that are still alive today. Many, in fact, are similar to the fish that currently inhabit the coral reefs of tropical oceans while others resemble species known from the open sea. Geopaleontological data suggest that the "Pesciara di Bolca" (the "Bolca Fishbowl") was a tropical lagoon separated from the sea by coral reefs and volcanic islands, analogous to what can be observed today in the Indo-Pacific.

The lignites and shales of two other deposits, Monte Purga and Monte Vegroni (Bolca)—both of which date to the Middle Eocene and are therefore slightly more recent than Pesciara—have yielded the fossils of crocodiles and marsh tortoises. What is of greatest interest for this article, though, is that they are also the source of spectacular fossil palms.

Among the various exposures in the Vicenza area that have yielded fossils of Lutetian-stage plants (Lower to Middle Eocene), the Fosse di Novale site near Valdagno is of special interest. There, where marls embedded in breccias and volcanic basalt sand crop out, the discovery of plant fossils suggests an island context. The dominant leaf types are lauroid (resembling trees in the laurel family) and leguminous (legumes, peas, or beans), which leads to the hypothesis that the paleoenvironment was quite warm and paratropical.

The Eocene of Northeast Italy concludes with the Croce dei Massignani exposure, where plant fossils likely date to the Lutetian stage (Lower Middle Eocene). The flora examined so far includes aquatic spermatophytes and fossils that can be traced to the living mangrove palm (*Nypa*).

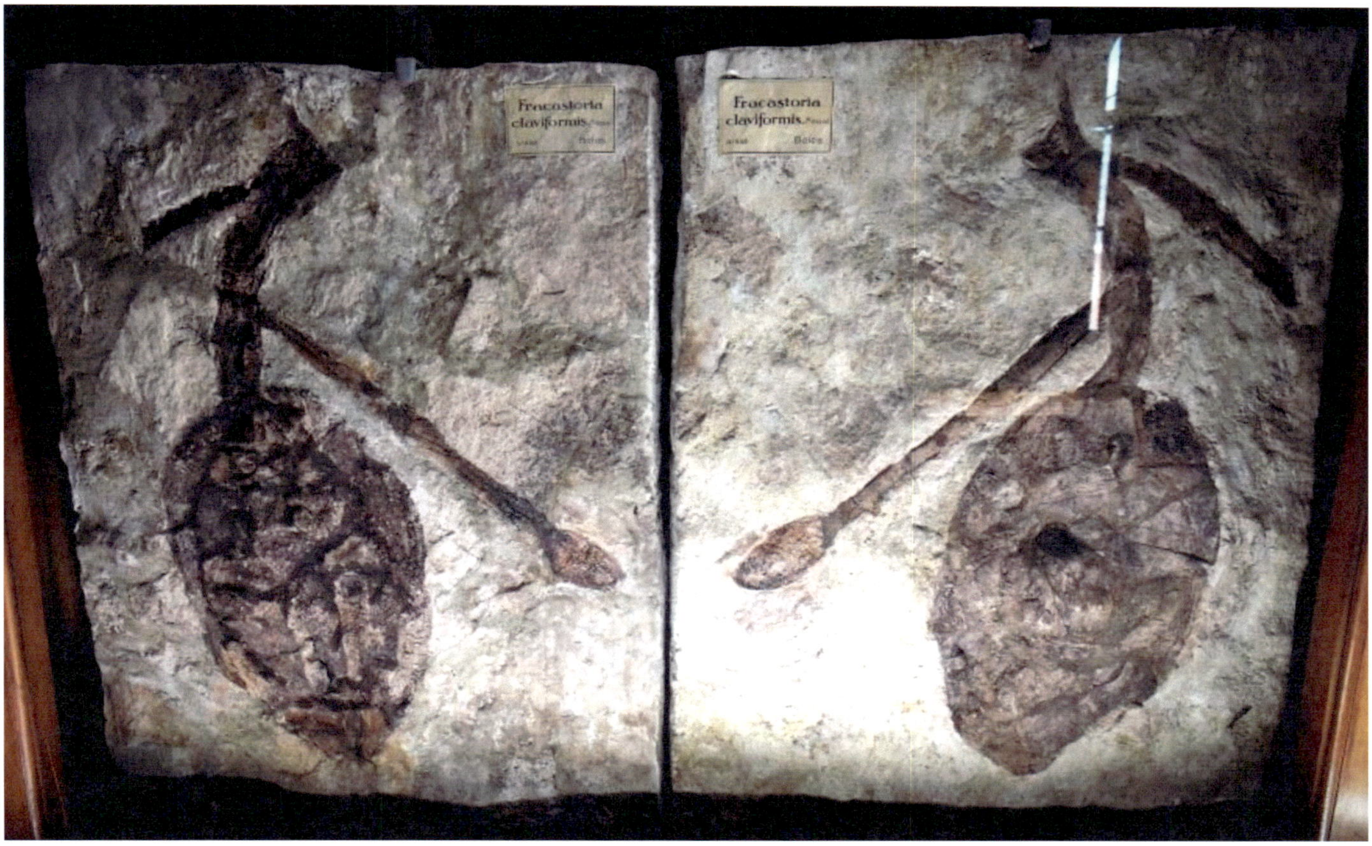

Above and below: *Large displays at the entrance to the Hall of Palms contain the fossilized fruit of primordial palms, many of which are similar to coconuts. Abramo Bartolommeo Massalongo, a paleobotanist born in the Province of Verona in 1824, named* Fracastoria obliqua (below) *and* Fracastoria claviformis (above) *in addition to some sixteen other species of* Fracastoria *and numerous other fossil plants. In both cases, the part and counterpart are shown.* Fracastoria *is now considered synonymous with* Nypa, *the mangrove palm.*

During the Oligocene, many sites in the Vicenza area yielded plant fossils, including the Montegioviale lignite deposit, the Salceto site and, in particular, the Chiavon Valley. The Chiavon leaf fossils are extremely well preserved and provide valuable evidence of Oligocene paleoflora that can be found in only a few other Italian exposures.

The Italian Neogene in the country's northeast is represented by specimens from the Miocene of Belluno (Province of Veneto), the Pliocene of Cornuda (Province of Treviso), and the Pleistocene of Belluno and of Leffe (Province of Bergamo). The climate of the period was tumultuous, marked by major environmental shifts that culminated in the Quaternary glaciations.

For any paleontology aficionado, a trip to Northeast Italy more than repays the effort. There, visitors can find not only the spectacular art and architecture of the Veneto but an exhibition of fossils that is objectively one of the most beautiful in the world. Giotto and his frescos revolutionized the history of painting, but the fossils in Hall of Palms, in their own way, revolutionized the history of paleontology.

Work Cited

Giusberti, L., Del Favero, L, and Roghi, G. (2014). [Chapter] 9. The Purga di Bolca-Vegroni Sites. Excursion Guidebook, Taphos 2014 Conference, *The Bolca Fossil-Lagerstätten: A Window into the Eocene World*, C.A. Papazzoni, L. Giusberti, G. Carnevale, G. Roghi, D. Bassi, and R. Zorzin, Eds. *Rendiconti della Societa Paleontologica Italiana* 4: 95-103.

Above: Nypa burtinii. *The imprint of a fruit from a plant related to the living "mangrove palm," which today grows in estuarine environments along the coasts of the Indian and Pacific Oceans. Lower Eocene, from the Pesciara di Bolca site. "Castellinia" (on the label) was included under* Nypa *by later paleontologists and is no longer used.*

THE HALL OF PALMS

Previous page, top: *Two of the large palm specimens on display near the entrance to the Hall of Palms.* Bottom left: *An exceptional specimen—an Oligocene palm germinating from a coconut-like seed. Part/counterpart. From the Valle del Chiavon site, Vicenza.* This page, above left: Sagenopteris sp. *A leaf of this Lower Jurassic pteridosperm from the Val d'Assa (Province of Vicenza).* Above right: Marzaria paroliniana, *another species first named and described by de Zigno. Lower Jurassic from the Zuliani Valley (Province of Verona).* Below: Voltzia haueri, *an extinct species related to modern conifers. Upper Triassic, Cave del Predil (Province of Udine).*

At left: Annularia sp., *a horsetail. Lower Carboniferous, Monte Corona (Province of Friuli).* Below, left: Sloanea peolai, *a leaf from a flowering Oligocene plant whose living descendants live in mostly tropical and subtropical zones. From the Valle del Chiavon site (Province of Vicenza). At right: Dolomite amber, among the oldest amber known, from the Upper Triassic. In Italy, it is mostly found around the town of Cortina d'Ampezzo (Province of Belluno). These 220-million-year-old droplets are known to contain bacteria, fungi, algae and protozoans that can be assigned to living genera.*

Paleontology NEW BOOKS *For Summer Reading*

Above: *One of Winifred Goldring's crinoids.* Next page, top right: *the women of* Daring to Dig. Bottom left*: A scene from the section devoted to the life of Annie Alexander, the discoverer of* Thalattosaurus alexandrae, *a Triassic marine reptile, and a foundiing patron of the University of California Museum of Paleontology.*

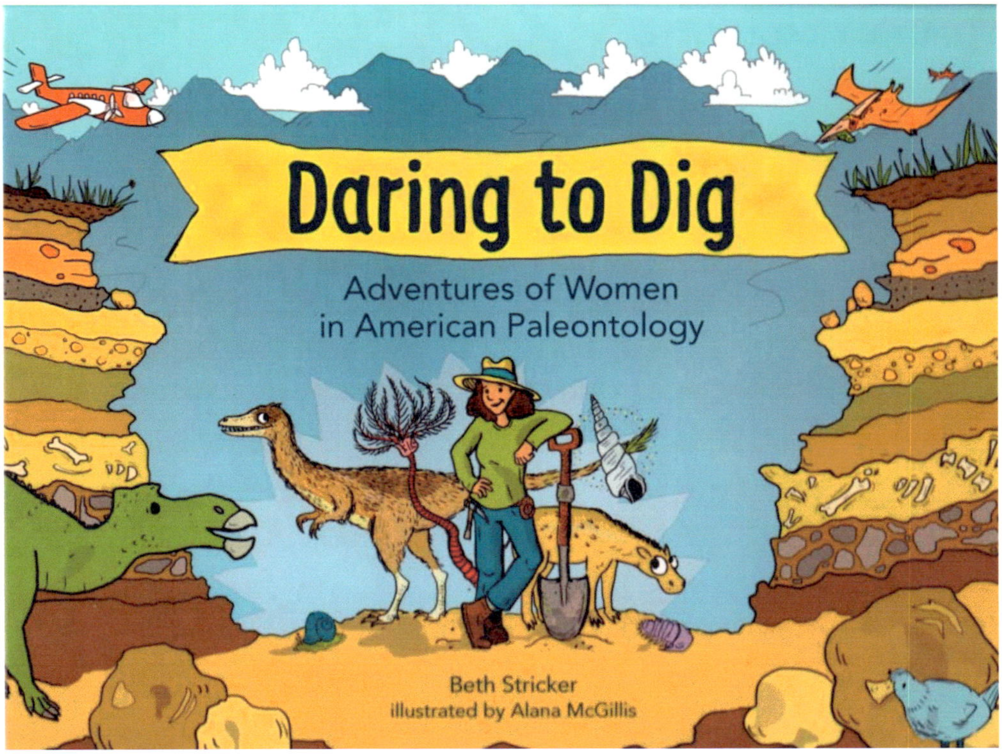

Daring to Dig: Adventures of Women in American Paleontology

Paleontological Research Institution ❖ ISBN-13: 978-0877105213

Great respect is due to Mary Anning, but if she's the only woman you can come up with in the history of paleontology, *Daring to Dig* will take you to school.

With text by Beth Stricker (Director of Exhibitions at the Paleontological Research Institution) and drawings by Alana McGillis (whose favorite paleontologist is Winifred Goldring "because she made pants for her friends"), *Daring to Dig* is aimed at kids, but there's a lot to love here for anyone interested in fossils and the people who study them for a living.

There's the prolific and daunting Winifred Goldring, who probably knew more than anyone ever has about the paleontology of New York—especially the Devonian and especially crinoids (she also really did make "bloomers" for her women scientist friends because they needed something comfortable to wear in the field); Karen Chin, an "alive and digging" coprolitologist (we made that word up) at the University of Colorado-Boulder; and Mignon Talbot, who discovered *Podokesaurus holyokensis*, a small Early Jurassic dinosaur that she named in part after Mount Holyoke College, where she was a professor of geology and geography (the whole name means "swift-footed lizard of Holyoke").

And that's just the beginning!

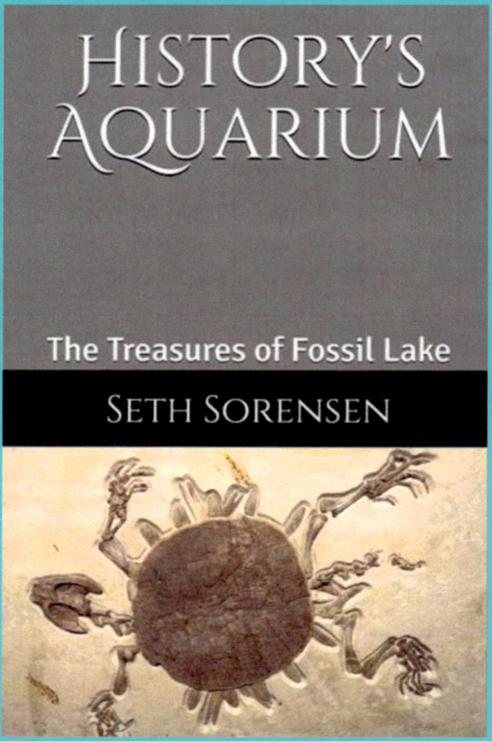

Part natural history of the phenomenon known as "Fossil Lake," one of three enormous tropical lakes that once existed in what is today Utah, Colorado, and Wyoming, part field guide for collectors and enthusiasts of Eocene-age Green River fossils, *History's Aquarium* packs a lot of information—and a lot of terrific photos—in only 125 pages. Author Seth Sorensen is the co-owner of the American Fossil Quarry, an historic quarry near Kemmerer, Wyoming. Known in times past by such names as Thompson Ranch and the South Dempsey Quarry, the American Fossil Quarry has, in its day, yielded—in addition to thousands of fish—a three-toed horse, several crocodiles, three turtles and, most extraordinarily, two primitive bats. Sorensen's envy-making tour of the exceptional fossils found in the quarry includes a rundown of the taphonomy of Fossil Lake—the process by which creatures that lived more than 50 million years ago became the well-preserved fossils that can be collected today—and an overview of fossil preparation and repair.

History's Aquarium: The Treasures of Fossil Lake
ISBN: 978-1980872122

Left: Heliobatis radians *and* Mioplosus. Above*:* Priscacara serrata. *Both from the "sandwich beds," American Fossil Quarry. Photos ©Seth Sorensen from* History's Aquarium.

"The discovery and notoriety of Fossil Lake in modern times happened much by accident. Although several early explorers in the west ... were aware of the fossils and deposits from Fossil Lake, there was not significant research and study of Green River Formation fossils until the late 1860s. It happened as part of the movement to expand transportation." — from *History's Aquarium*.

Next up in Siri Scientific Press's series of authoritative monographs in paleontology is *Fossils of the Milwaukee Formation*, a detailed and lavishly illustrated exploration of the fossil fauna of little known Middle Devonian deposits found only in a small area in southeast Wisconsin. Kenneth Gass, author of *Solving the Mystery of the First Animals on Land: The Fossils of Blackberry Hill*, a study of the Cambrian trace fossils of central Wisconsin, is joined by three prolific writers and scientists in this effort to document accurately and completely a beautiful and intriguing Devonian biota.

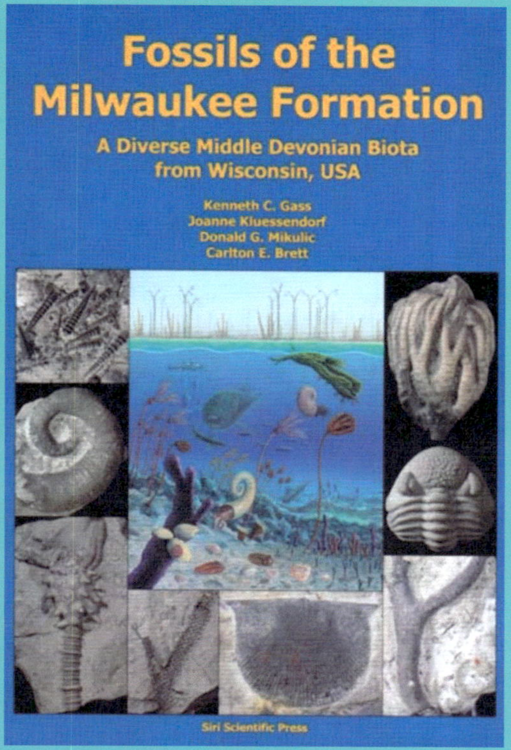

Fossils of the Milwaukee Formation: A Diverse Middle Devonian Biota from Wisconsin, USA

Siri Scientific Press ❖ ISBN-13: 978-0-9957496-7-2

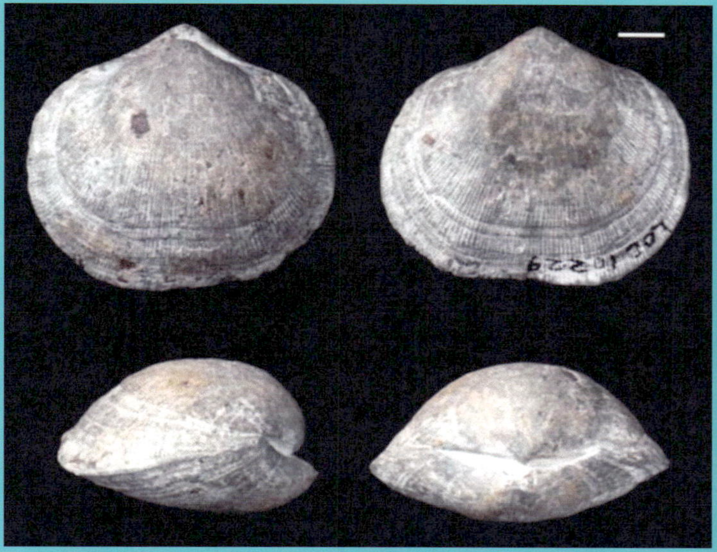

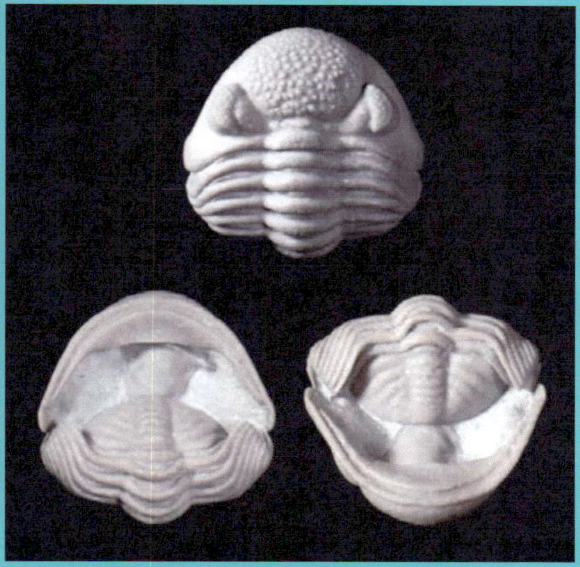

Three of the remarkable photographs in Fossils of the Milwaukee Formation. Top of page: *Acleistoceras? calvini (Cleland, 1907), an oncocerid cephalopod; scale bar = 1 cm.* Above left: Schizophoria iowensis muscolosa *Griesemer, 1965, an orthid brachiopod; scale bar = 5 mm.* Right: *Three views of a large, partially enrolled* Eldredgeops? norwoodensis *(Stumm, 1953). Note detached hypostome in the bottom left image and, at bottom right, the pygidium and a partially detached thoracic segment; scale bar = 5 mm.*

"In the mid-1840s Increase A. Lapham was the first person on record to have collected fossils from what would later be known as the Milwaukee Formation.... Lapham recognized the similarity of this Devonian rock to that used in the natural cement industry.... In 1875, Joseph R. Berthelet established the Milwaukee Cement Company and, by 1891, the MCC was the largest producer of natural cement in the country.... During this time, conditions could not have been better for discovering fossils in the Milwaukee Formation, and "gentlemen paleontologists" would pay the quarry workers generously for their best fossil finds." —from the Introduction, *Fossils of the Milwaukee Formation.*

Enrico
BONINO

An excerpt from the new book, not yet available in English, *Spiagge Cambriane: Meduse e Tappeti Algali*

We know them for the nearly hypnotic undulation of their long tentacles (death traps for the incautious organisms that blunder into them) and the rhythmic and voluptuous contractions of their umbrella-like bells. Jellyfish exist in a marine environment that is nearly inaccessible to human beings but which gently transports them all over the globe.

These alien, nearly supernatural creatures immediately take us to far-off eons and distant worlds—in part because they have for so long been considered the first "complex" organisms to appear on Earth and have been called the first important ocean colonizers.

In fact, however, their meandering presence in the oceans of the Earth is relatively recent in the long history of the planet: No certain evidence is known of their existence before the Cambrian. In fact, the first verified medusozoan fossils are preserved in the Lower Cambrian sediments (approximately 520 million years old) of the Zabriskie Quartzite (a formation in the northern Mojave Desert of California and Nevada). There, a series of disc-shaped, sometimes misshapen macrofossils have been found, though their placement in a more specific taxonomic group than Phylum Cnidaria has so far proved impossible, and their affinities are constantly revised. [Ed. note: Fossil structures that could be interpreted as jellyfish are classified scientifically as medusozoans so as not to create confusion among superficially similar creatures, all of which might colloquially be referred to as "jellyfish." Medusozoans include the Classes Cubozoa (box jellyfish), Hydrozoa (hydrozoans, polyps, and the Portuguese man o' war), Staurozoa (stalked jellyfish), and Scyphozoa

"*Noah's Ravens* is truly a masterpiece for the science of fossil trackmakers. . . . This volume will soon be a standard within (ichno-) vertebrate paleontology and a role model for subsequent books."
—Annette Richter, Lower Saxony State Museum at Hannover, Germany

"Armed only with clay, plaster and paint, Charles R. Knight brought the dry bones back to life, to live on in our memories and dreams—and sometimes our nightmares as well."
—Richard Milner, author of *Charles R. Knight: The Artist Who Saw Through Time*

"You will not be disappointed!"
—*Journal of Vertebrate Paleontology*

Revised, updated, and expanded, the second edition of *Oceans of Kansas* adds new twists to the fascinating story of the vast inland sea that engulfed central North America during the Age of Dinosaurs.

Where Dinosaurs Roamed paints a picture of the incredible diversity of plant and animal life in the Late Cretaceous and of the fossil record left behind in the Grand Staircase region in southern Utah.

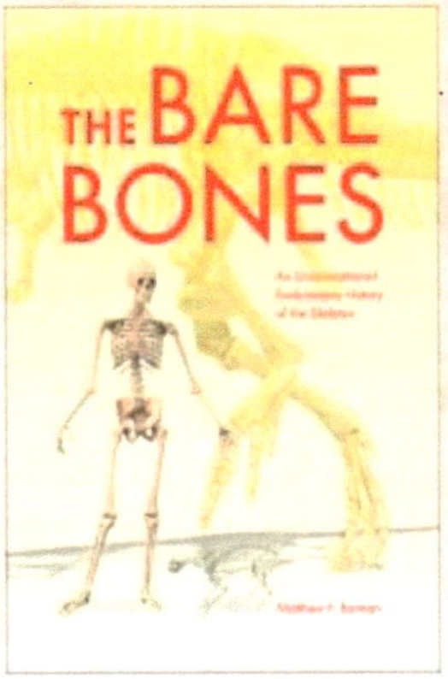

"No bones about it, a text like *The Bare Bones* was sorely needed in the popular literature of vertebrate paleontology. . . . It is written in an enjoyable and readable fashion that will absolutely delight all sorts of readers."
—*Palaeontologia Electronica*

"Well designed, handsome and fantastically well edited, congratulations are deserved to the editors for pulling together a vast amount of content, and doing it well." —Darren Naish, *Tetrapod Zoology*

"This handsome book, written by a leading expert in South American paleontology, is profusely illustrated with maps, time charts, color photographs of fossils, and exquisite life reconstructions."
—*Choice*

"This book is a highly enjoyable romp through the intellectual history of evolutionary biology, using turtle evolution as its red thread."
—*Natural History Book Service*

INDIANA UNIVERSITY PRESS

Explore Your World
iupress.indiana.edu

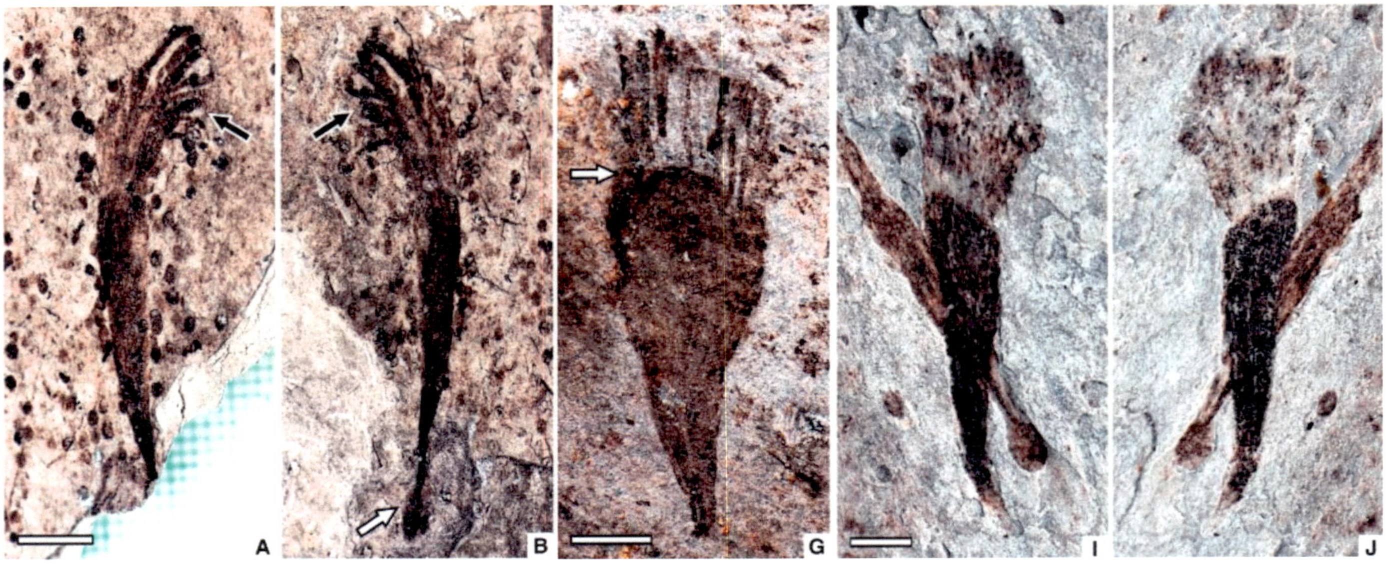

Above*: These images, taken from Wan et al. (2016), illustrate Ediacaran fossils from the Lantian Biota in South China that the authors identified as a new species,* Lantianella laevis. *A) Part and counterpart of the holotype, showing tentacle-like structures (black arrows) and, in B), a spherical holdfast (white arrow). G) Another specimen shows the margin of the cone-shaped body (white arrow); I & J) Part/counterpart of amother specimen showing the sharp contrast between the lighter-colored crown of tentacles and the darker conical body; the images also show the holdfast structure. Wan et al. interpreted* Lantianella *as a conularid fossil possibly related to scyphozoan cnidarians. All scale bars represent 5 mm. Images © the authors.* Below: *Specimens of genus* Cambrorhytium *from the Lower Cambrian Burgess Shale, Walcott Quarry, British Columbia, Canada. Cambrorhytium is an enigmatic fossil genus also known from California and China. Conical and tube-like, it has been interpreted as a cnidarian polyp (Class Staurozoa). Images from* Spiagge Cambriane.

(the "true" jellyfish). Keep in mind, too, that while all "jellyfish" are cnidarians, not all cnidarians are jellyfish. Corals and sea anemones are also cnidarians, for example. Comb jellies, just to complicate matters, are placed a separate phylum: Ctenophora.]

In the second half of the 1940s, the phenomenal Ediacaran fossils of Australia (which date to 635-541 million years ago) came to light. Among the many other intriguing fossils found in those exposures were disc-shaped specimens that were initially interpreted (wrongly) as medusozoans. Subsequently, many other circular fossils were also described as fossil jellyfish. In fact, we now know that many of these enigmatic fossils represent petal-shaped attachment structures, anemones, other little-understood organisms such as eldonioids, ichnofossils (resting traces, e.g.), burrows, sand volcanoes, or the marks left by accumulations of gas either just below or just above the algal mat.

Still, organisms did appear in the distant and ancient Ediacaran Period that may have belonged to Phylum Cnidaria. Some of these, whose fossils have been found in China, were generally conical in shape and were equipped with what appear to be tentacles or similar structures; other aspects of these fossils are more difficult to interpret. One hypothesis is that they are related to the conularids, a group that has generally been thought to have cnidarian affinities, though this connection remains speculative.

At least in theory, then, the origins of the Cnidaria might be traced to the base of Ediacaran, in a period that followed the end of the great Marinoan glaciation. After the retreat of this worldwide glaciation (sometimes called "Snowball Earth"), a process began that ushered in the proliferation and diversification of many forms of life. The flora and fauna of the Lantian Biota in China are evidence of this dramatic burst of development; there, fossils such as the conical specimens described above are preserved as carbonaceous films in black shales.

Fossil Jellyfish

The question then arises naturally: How in the world could an organism composed almost entirely of water be preserved as a fossil, especially in a marine environment that

At right: *Three examples of disc-shaped fossils originally (and apparently incorrectly) interpreted as cnidarians. E)* Tirasiana disciformis, *now understood as the holdfast of a frond-like organism. A)* Aspidella terranovica, *which Gehling et al. (2000) described as having been "reinterpreted perhaps more than any other Precambrian taxon ... variously regarded as a fossil mollusc or 'medusoid,' a gas-escape structure, a concretion, or a mechanical suction mark." They cautiously interpret it, like* Tirasiana, *as the attachment structure of a frond-like organism. B)* Cyclomedusa delicata, *a widely distributed Ediacaran fossil, now believed to be as a microbial colony. All three specimens are from the White Sea, Russia. Images from* Spiagge Cambriane; *used by permission.*

Fossilization of a jellyfish may begin with an unusual tide, a bloom, or a storm. A) An organism is transported to the intertidal zone—the area between high and low tides—a harsh and unforgiving habitat that is shaped by forces of both the sea and the land. There it becomes trapped in a pool of water and can no longer return to the open sea. B) In its effort to escape, the jellyfish expands and contracts its bell and, in the process, ingests quantities of sand and silt. C) The creature's pulsations also create round, concentric imprints that can be seen in the subsequent phases as the animal begins to shrink from desiccation. If no other organisms appear to feed on the carcass, and if an algal mat is present, the initial conditions for fossilization are set in place. D) Later, tides and winds transport sand and silt that cover the carcass, initiating the deposition of sediments that may, in time, preserve a fossil. Reproduced from Spiagge Cambriane; *used by permission.*

Above: *Clues to the presence of algal mats in the Jurassic Solnhofen basins. The strips visible in these two plates were probably caused by bottom currents that rolled the algal mat like a carpet. Both are from a site near Mörnsheim, Eichstätt District, Bavaria, Germany. Scale bar is 10 cm.* Below: *A recreation of a shallow-water Middle Pennsylvanian Mazon Creek environment, showing medusozoans above and on an algal mat. Represented are the Scyphozoans* Octomedusa, Essexella, Reticulomedusa, *and* Lascoa, *and the Cubozoan* Anthracomedusa. *At right are a pair of "Tully Monsters,"* Tullimonstrum gregarium, *a still enigmatic organism so far known only from the Mazon Creek, Illinois Lagerstätte. All images taken from* Spiagge Cambriane; *used by permission.*

was in continuous flux? In fact, though jellyfish are abundantly distributed in all of today's oceans, their presence in the fossil record is extremely unusual. They've been discovered only in exposures that have yielded exceptionally well preserved fossils and which are associated principally with lagoon deposits and beaches (very rarely in deep water), a consequence of the jellyfish's density—which means it is likely to float and be transported by currents.

Given the general rarity with which taxa without a skeleton or other hard parts are fossilized, the discovery of fossils of soft-bodied organisms is hugely significant from an evolutionary, taphonomic, and paleoenvironmental perspective. The scarcity of jellyfish or other medusozoan fossils isn't solely the result of the fact that these organisms are gelatinous and, as a result, unlikely to be preserved. It's also influenced by the presence or absence of predation, bioturbation, and physical, chemical, and mechanical phenomena that cause the decay and disappearance of almost all carcasses that wash up on the shore.

The large number of specimens collected, for example, in the Zabriskie Quartzite, Potsdam Sandstones (Northern New York), and the Elk Mound Group (Wisconsin), suggest that the mechanisms that reached their apex during the Cambrian were similar to those we can observe today in modern coastal environments. The frequency of jellyfish Lagerstätten in the Lower Paleozoic, then, testifies to an absence of predation, bioturbation, and feeding on the beaches of the ancient continental margins. What is more, the distinctive paleoecological conditions of these environments were highly significant to the fossilization process because they promoted the development of algal colonies (as described more fully below). That, in turn, facilitated the preservation of the most delicate structures of many soft-bodied creatures.

The process of "terrestrialization" by the first organisms who abandoned the Cambrian oceans and began the slow, inexorable colonization of dry land included arthropods, mollusks, worms, and others, and traces of their ancient movements are magnificently captured in the sandstones laid down in ancient coastal areas, where conditions allowed their preservation as fossils. These same ocean/continent interfaces tell another story, too: a record of major jellyfish strandings.

Jellyfish strandings take place as the result of storms or unusual tides and may involve hundreds or even thousands of individuals. They may also be the result of "blooms" (periods of a significant, usually sudden population increase) caused by higher concentrations of nutrients in the water, warmer temperatures, the absence of predators, and other factors. Some of these stranding events produced fossils.

Algae

The first organisms to appear on the Earth were single-celled algae, which massively colonized the oceans, the continental platforms and, presumably, even the continents themselves where hot springs existed and where chemical and temperature conditions were sometimes extreme (from which comes the term "extremophile").

These strikingly diverse habitats equipped algae with a radical ability to adapt to a wide variety of environmental conditions. Initially, algae were organized into simple mats that grew largely in horizontal layers in environments marked by relatively calm waters.

Later, the stromatolites (from the Greek *stromatos* or "level, strarum" and *lithos* or "rock") appeared and, in varied and intricate forms, survived for nearly three billion years. They represented an especially successful adaptation to environmental dynamics: the formation of true communities in which numerous species of microscopic algae worked together for mutual survival and to create highly complex bioconstructions. Oceans, coasts, beaches, and tidal zones were consequently colonized by algal mats. These colonies were able to thrive because of the lack of detrivorous or grazing organisms along the shoreline and the absence of bioturbation processes that would have mixed surface sediments and interrupted the colonies' development.

The continents were completely barren but, wherever water existed, colorful algae developed; their ancestors can be seen today in especially hostile environments such as thermal vents, hot springs, or hypersaline zones where life would be difficult and demanding for larger organisms.

Virtually all the exposures that have yielded the fossils of soft-bodied organisms have shown direct or indirect evidence of the presence of colonies of algal mats that covered the depositional surface.

Virtually all exposures that have yielded the fossils of soft-bodied organisms have shown evidence of algal mats that covered the depositional surface.

Hand-Crafted Display Boxes
Made to order in Black Walnut, ideal for arrowheads, small fossils, coins & other collectables.
Sizes
9" x 12" ~ $85
12" x 18" ~ $95
RICK HEBDON ❖ Kemmerer, Wyoming ❖ warfosq@silverstar.com ❖ 307.883.2445

Fossil Shack
The World's Most Trusted Name in Fossils
Fossil Shack can meet your fossil needs in a wide variety of areas including:
*Fossil Preparation Supplies (Air Scribes, Pin Vices, Glue, etc...)
*Fossil Repair and Restoration Supplies (PVA, Epoxy Putty, Debonder)
*Full Dinosaur, Pterosaur, Marine Reptile and Mammal skeletons
*Incredible Green River Fossils
*Fossil Home Decor & Fun Fossil Swag
*Meteorites, Minerals, Gems & More
From Dinosaurs to Mammoths and Fossil Prep Supplies to Meteorites, we have what you want!
www.FossilShack.com

Cyanobacteria

Cyanobacteria are known to have been the first organisms to be able to convert CO_2 into oxygen and to release that oxygen into the primordial atmosphere. As a result, algal mats composed of cyanobacteria played a significant role in the preservation of soft tissues, prior to and during the process of fossilization. They ensured that the decomposition of organic material would take place as slowly as possible, protecting soft tissues from predation and from mechanical processes such as waves and currents. Algal mats also created special macro- and microenvironmental conditions in which the concentration of oxygen was low. That, in turn, promoted an especially rapid precipitation of minerals, and the quick substitution of organic matter with clay and sulfides fostered the preservation of organisms locked in the sediment.

Cyanobacteria are found on the upper part of the algal mat because, as organisms with an aerobic metabolism, they need light to activate photosynthesis. They may be covered by sand or slime and contain pigments such as scytonemin. This yellowish-brown substance is triggered by exposure to ultraviolet radiation (UV light) and is found on the exterior of some cyanobacteria, where it forms a kind of protective "shield."

Cyanobacteria with a similar sort of protection have been found in stromatolites that date to the Proterozoic (2.5 billion years ago). At that time, Earth's young atmosphere was unable to provide much protection from the Sun's ultraviolet rays, which were consequently much more intense than they are today. Presumably, then, the development of scytonemin played a vital role in the evolution of these algae in subtidal environments. It was there where they were more likely to be exposed to damaging UV rays, and where a filtering mechanism became essential to life.

Works Cited

Gehling, James G., Narbonne, Guy M., and Anderson, Michael M. (2000). The First Named Ediacaran Body Fossil, *Aspidella terranovica*. *Palaeontology 43* (Part 3): 427-456.

Wan, B., Yuan, X., Chen, Z., Guan, C., Pang, K., Tang, Q. and Xiao, S. (2016). Systematic Description of Putative Animal Fossils from the Early Ediacaran Lantian Formation of South China. *Palaeontology 59*: 515-532. doi:10.1111/pala.12242.

Made in the USA
Las Vegas, NV
09 November 2026

57764359R00031